Know Your Sheep

Jack Byard

Old Pond Publishing

First published 2008, reprinted 2008 (four times), 2009 (twice)

ISBN 978-1-905523-83-2

Published by
Old Pond Publishing Ltd
Dencora Business Centre
36 White House Road
Ipswich IP1 5LT
United Kingdom

www.oldpond.com

Book design by Liz Whatling
Printed and bound in China

Contents

Acknowledgements

I am indebted to many sheep breeders for their help in compiling this book. I am particularly grateful for the help of the following: the North Yorkshire Smallholders Society and its members; Moira Linaker (the Eden flock); Patrick Carr, Robert E Rose, John Terry and www.north-country.co.uk.

Picture Credits

Plate (1) Robert Carroll, *(2)* Catherine A MacGregor www.macgregorphotograpy.com, *(3)* claudpettis@yahoo.com, *(4)* Andrew Morton Jnr, *(5)* The Society of Border Leicester Sheep Breeders, *(6)* Cheviot Sheep Society of NZ Inc., *(7)* The Damburgh Flock, *(8)* Dalesbred Sheepbreeders Association Ltd, *(9)* The Dartmoor Sheep Breeders Association, *(10)* John Terry, *(11)* Phil Dallyn, Bulleigh Park Devon Closewool, *(12)* Devon & Cornwall Longwool Flockbook Association, *(13)* Dorset Down Sheep Breeders Association, (14) Dorset Horn and Polled Dorset Sheep Breeders Association, *(15)* Exmoor Horn Sheep Breeders Society, *(16)* Hampshire Sheep Breeders Association, *(17)* John Marriott, *(18)* Shetland Sheep Breeders Group, *(19)* Rosewood Farms, *(20)* Leicester Longwool Sheepbreeders Association, *(21)* Lincoln Longwool Sheepbreeders Association, *(22)* www.baylham-house-farm.co.uk, *(23)* Lleyn Sheep Society, *(24)* A Dixon, *(25)* Masham Sheep Breeders Association, *(26)* North Country Cheviot Breeders Association, *(27)* Oxford Down Sheep Breeders Association, *(28)* Seven Sisters Sheep Centre, *(29)* American Romney Breeders Association, *(30)* Rough Fell Breeders Association, *(31)* The Eden Flock, *(32)* Shetland Breeders Group, *(33)* Shropshire Sheep Breeders Association, *(34)* Clive Pritchard, *(35)* The Suffolk Sheep Society, *(36)* North York Moors National Park Authority, *(37)* Teeswater Sheep Breeders Association, *(38)* & *(39)* www.meadsheep.co.uk, *(40)* The Wensleydale Longwool Breeders Association, *(41)* Whiteface Dartmoor Sheep Breeders Association.

Foreword

My hope is that this book will increase your enjoyment of a day in the country.

The idea for it came after I talked to a group of eight- and ten-year-old children who believed that a sheep was a sheep and there was only one breed. When I told them that there were dozens of different breeds they thought this was 'cool'.

I started to gather more detailed information and asked local children to help with the research. Some of the girls, who came from farming backgrounds, both lowland and hill, were able to add 'professional' touches. Their parents were most enthusiastic about the project because they thought the general public should know as much about farming as possible, especially in view of the disastrous consequences of the foot and mouth outbreaks and other problems facing the farming community.

The sheep breeds I have chosen for this book are, in my opinion, the ones you are most likely to see on your rambles. Wherever possible I have chosen photographs showing the animals in their natural surroundings.

JACK BYARD

Bradford, 2008

Beulah Speckled Face

Location

The British Isles: originally only in Wales but now country-wide. There are now small flocks across the British Isles and also in Europe.

Description

Larger than most mountain sheep. The face and legs are distinctly speckled and free from wool. The ears are erect and pointing slightly forward. The rams and ewes are without horns. The wool is white.

The full title of the breed is the Eppynt Hill and Beulah Speckled Face. The breed has been seen on the Welsh hills for over a hundred years without the introduction of female stock. The high-quality wool is used for soft-handling tweed, fine flannels and knitting wools; the coarser wool is used in the mattress, rug and carpet trade. The Beulah Speckled Face also produces top-quality meat.

Blackface

Location

The British Isles: originally the Scottish borders but now country-wide – the most numerous breed in Britain. Also found in Europe, North America and South America.

Description

Black or black-and-white face and legs. The rams and ewes have horns.

The breed's origin goes back to at least the twelfth century when monastery records mention Blackface or Dun. In Scotland the Blackface is known as 'Blackie'.

The finest quality Wilton and Axminster carpets are made from Blackface wool. Scottish and Irish tweeds are also produced from this wool, Harris and Donegal possibly being the best-known examples. The strong, coarse wool varieties are sold to the Italian market for the mattress trade. The Blackface also produces top-quality meat.

3.

Black Welsh Mountain

(Defaid Duon Mynydd Cymreig)

Location

British Isles and North America

Description

A small black sheep with no wool on the face or on the legs below the knees. The ewes are hornless.

Writings in the middle ages frequently mention the black mountain sheep. This is the only completely black sheep in the United Kingdom. When the tips of the wool are bleached by the sun they turn a beautiful reddish-brown. Lambing usually takes place from mid-February to the end of March. The wool is used undyed for many cloths and is mixed with whites and greys for the check tweed produced in Wales. The Black Welsh Mountain also produces top-quality meat

Bluefaced Leicester

Description

A broad muzzle and a tendency towards a Roman nose. On the head the colour of the skin should show blue through the white hair. The body wool is tightly purled (like small knots). There is no wool on the head and legs. The rams and ewes are without horns.

The breed was developed during the late 1800s and early 1900s near Hexham, Northumberland. By 1930 it was recognised as a distinct breed. Crossing the Bluefaced Leicester with pure hill breeds, especially a Swaledale or Cheviot, produces 'mule' lambs that are able to withstand quite harsh weather. The wool is the finest of any native breed and is blended with silk and mohair in the production of high-quality fashion garments; it is also used for moquettes and hosiery and is greatly prized by home-spinners. The Bluefaced Leicester also produces quality meat.

Location

British Isles, Australia, New Zealand and North America.

Border Leicester

Location

British Isles, Europe, Australia, New Zealand and North America.

Description

A large sheep that looks proud and graceful with its prominent erect ears and a Roman nose. There is no wool on the head or below the knees. The rams and ewes are without horns. The wool is white and dense.

There is no doubt about the origins of this breed: the sheep are direct descendants of the Dishley Leicester bred by Robert Bakewell (1725-95) of Dishley, Leicestershire. The breed was introduced into Northumberland in 1767 and was soon established on both sides of the border. The wool is used for dress fabrics, lining materials, hosiery and hand knitting wools. The Border Leicester also produces top-quality meat.

6.

Cheviot

Description

The Cheviot has fine white hair on its face, head and legs. The ears are erect with a ruff of wool behind them. The back is broad and the rams may have horns.

These white-faced sheep have been run in the borders for several hundred years. The Cheviot sheep can be found up to 3,000 feet above sea level and are expected to survive on the hills throughout the year. They are fed hay in extreme conditions and prior to lambing. The wool is used for lightweight suitings, tweeds, knitting wool, hosiery, blankets and rugs. A small quantity is used by hand-spinners. The Cheviot also produces quality meat.

Location

British Isles: originally the Scottish borders. Also to be found in Australia, New Zealand and North America.

Clun Forest

Description

Tan to dark brown face and legs, the forelock of white wool is almost crew cut. There is little or no wool on the legs. The rams and ewes are without horns

The Clun Forest breed originated in the hilly areas of south-west Shropshire on the Welsh borders and took its name from the ancient market town of Clun. It has been suggested that semi-nomadic shepherds who lived in the forest nearly a thousand years ago bred the original Clun sheep. Whilst they may live at altitudes of 1,500 feet above sea level, others are content on rich lowland pastures. Longevity is a strong point of this breed with records showing ewes can still be breeding at twelve years of age. The ewes are excellent mothers and produce plenty of rich milk. The fine, dense wool is used for woollen cloths, fine hosiery, knitting wools and industrial felts. The Clun Forest also produces quality meat.

Location

British Isles: originally in the Welsh borders, but now across Great Britain. The breed is also found in Europe and North America.

Dalesbred

Description

A black face with a white spot mark on each side of the nose, with the nose end becoming grey. The legs also have distinctive black-and-white markings; there is no wool on the face or legs. Rams and ewes have round, low-set horns.

It was in 1930 that breeders in the western dales of Yorkshire formed an association to fix the type of sheep they had been breeding for generations. These sheep are one of the possible maternal parents of the Masham breed and mule sheep which are the mainstay of lowland sheep production. The Dalesbred are hill sheep that are very adaptable to climatic changes and can survive the bleakest conditions. The wool is used for tweeds and carpets. The Dalesbred also produces quality meat.

Location

British Isles: upper Wharfedale, the central Pennines and the part of Cumbria that was originally Westmorland.

Dartmoor

Description

Nostrils mottled or spotted with black or grey. The head and body are well covered in wool, reminiscent of an old English sheepdog. The rams and ewes are without horns.

The Dartmoor, also called the Greyface Dartmoor, descended from the sheep which grazed around the moor, improved by crossing with other local breeds. The breed has been in existence for over a century. The Dartmoor has a superb constitution, enabling it to survive the harshest of winters. The wool is used for blankets, carpets and wool cloths. The Dartmoor also produces top-quality meat.

Location

British Isles

Derbyshire Gritstone

Location

British Isles: mainly the Peak district of Derbyshire and the Pennines of Yorkshire and Lancashire. It is also found in Wales.

Description

One of the largest mountain and hill sheep. The face and legs, which are free of wool, have clean-cut black-and-white markings. The rams and ewes are without horns.

The Derbyshire Gritstone, one of the oldest British breeds, originated in the Derbyshire Peak District in the mid eighteenth century. It is an adaptable breed well able to survive the harsh conditions up to 2,000 feet above sea level. The wool is used for high-quality hosiery, knitted outerwear and underwear. The Derbyshire Gritstone also produces top-quality meat.

Devon Closewool

Location

British Isles: mainly south-west England, Exmoor and the surrounding countryside.

Description

A white face and short ears. The fleece is very dense, and the face and legs are covered with short, white hair.

Originating from the cross of Exmoor Horn and Devon Longwool, the breed has existed for over a hundred years. The Devon Closewool Society was established in 1923. The dense coat gives excellent protection in harsh weather so that the sheep can survive these conditions with little or no help. The wool is used to enhance the rugged appearance of many tweeds and is also used for hosiery fabrics. The Devon Closewool produces quality meat.

Devon and Cornwall Longwool

Description

A white face with a black nose. The head and body are well covered with long, curly, white wool. The rams and ewes are without horns.

Developed from two of the oldest West Country breeds, this grassland sheep produces one of the heaviest fleeces in Great Britain. The lamb's wool is prized for its use in knitwear, flannels and dress fabrics. The adult wool is very strong, so it is used for tweeds, carpets and rugs. These sheep also produce quality meat.

Location

British Isles: mainly south-west England.

Dorset Down

Description

The Dorset Down has a deep chest. Its face, ears and legs are brown. The rams and ewes are without horns.

The Dorset Down is a result of work carried out two hundred years ago by Homer Saunders of Watercombe and a Mr Humfrey of Chaddleworth who improved the local Down breeds. The wool is of the lighter grade. It is used for flannel, dress material and speciality knitting yarns. The Dorset Down also produces delicately flavoured meat.

Location

British Isles: mainly the south-east, south-west, west Midlands and Wales. Also in Europe, New Zealand and Australia.

Dorset Horn and Poll Dorset

Description

The Dorset Horn has a white face with pink nose and lips. The legs and ears are also white. The horns are thin and symmetrically curved in both sexes. The Poll Dorset is identical apart from being without horns (the meaning of 'polled').

The Dorset Horn breed was established in 1892, the Poll Dorset originating in Australia. Both breeds are able to breed at any time of the year and to produce one of the highest quality wools in the country. This fine-quality wool is used for dress fabrics, flannel, hosiery, fine tweeds and speciality knitting yarns. The Dorset Horn and Poll Dorset also produce quality meat.

Location

British Isles and worldwide.

Exmoor Horn

Description

A compact, chubby-looking sheep with a white face and black nostrils. The legs are covered with white wool. The rams and ewes have horns.

The Exmoor Horn is an ancient breed which has existed since time immemorial. The breed is able to survive the harsh winters on Exmoor and the bleak Brendon Hills. The Exmoor Horn was a major influence on the West Country's reputation for quality wool. The area is famous for special wool cloths of superb quality and finish. Other uses are hosiery, felts, knitting yarns and tweeds. The Exmoor Horn also produces quality meat.

Location

British Isles: mainly upland areas of south-west England.

Hampshire Down

Description

A rich, dark brown head, legs, face and ears. The head, forehead and cheeks are covered with wool. The legs are powerful and set wide apart. The rams and ewes are without horns.

The Hampshire Down was established over 150 years ago, and the lambs are noted for being robust and hardy. The main economic advantage of Hampshire Down sheep is that they grow more rapidly than sheep of other breeds. The fine, dense wool is used for hand-knitting wools, hosiery, felts and flannels. The Hampshire Down also produces top-quality meat.

Location

British Isles: originally the southern counties but now nationwide and throughout the world.

Herdwick

Location

British Isles: mainly the Lake District fells.

Description

White or grey head, face and legs. The ears are white and the legs are covered with short, strong hair. The lambs are born with black wool which turns blue-grey with age. The rams have creamy-white, curved horns; the ewes are without horns.

This ancient breed originated in the Lake District. Its breed points were agreed in the 1840s. One remarkable instinct of the Herdwick is that it will never wander far from where it was born. The sheep of this breed live their entire lives in the mountains with no supplementary feeding. The wool is mainly used for carpets. Undyed wool from all age-groups of the Herdwick is used for speciality fabrics and knitting wools. The Herdwick also produces top-quality meat.

Jacob

Description

Brown-and-white or black-and-white patches. The rams and ewes can have two or four horns.

The origin of the Jacob, also known as the Spanish sheep, is unclear. However, in Genesis 30 there is an account of Jacob selectively breeding spotted sheep. The pattern and colour are thought to have originated in Syria three thousand years ago and moved through Sicily and Spain. Jacobs were introduced into Great Britain in the eighteenth century as ornamental sheep in parks. The Jacob Society was formed in 1969. The wool is much in demand by hand-spinners. The Jacob also produces quality meat with a sweet flavour.

Location

British Isles: country-wide and also in North America.

Kerry Hill

Description

Well-defined black-and-white markings on the face and legs, and a black nose. No wool on the head and legs. The rams and ewes are without horns.

The breed is named after the village where it originated – Kerry in Wales. The earliest mention of the breed was in 1809. The distinctive black-and-white markings make the Kerry Hill easily recognisable. The wool is amongst the softest in Great Britain. It is used for tweeds, flannel, knitwear and furnishing fabrics, and is especially suitable for deep-textured designs. The Kerry Hill also produces top-quality meat.

Location

British Isles: mainly the Welsh borders and Midlands. Also in the Netherlands.

Leicester Longwool

Location

British Isles: originally north-east England. Now in Australia, New Zealand and North America.

Description

A large sheep with curly wool and a white face, black nostrils and black lips. The head is well covered in wool and the legs are white. The ears are fairly large and occasionally have black spots. The rams and ewes are without horns.

Today's Leicester Longwool is a direct descendant of a breed developed by Robert Bakewell more than two hundred years ago. The breed's hardiness ensures that it survives the rigours of the climate. The wool is used for suit linings, coating cloths, knitting, rug-making, hand-spinning, tapestry, wall hangings and soft furnishings. The Leicester Longwool also produces quality meat.

Lincoln Longwool

Location

British Isles: mainly Lincolnshire but dotted around the country. It is also found in North America, Australia and New Zealand.

Description

The largest of the longwool sheep. A white face and head with a woolly forelock, with dark ears pointing slightly forward. The rams and ewes have no horns.

Over many hundreds of years the Lincoln Longwool has been developed to produce high-quality, strong, lustrous wool to make hard-wearing, warm, durable cloth. A society to safeguard the breed was formed in 1796. The wool has an extremely diverse range of uses, including quality suitings, upholstery, braids, linings, carpets, dolls' wigs, plush soft toys and weaving. Hand-spinners frequently mix the wool with mohair to produce a 'fluffy' yarn. The Lincoln also produces top-quality meat.

Llanwenog

Location

British Isles: mainly Wales, but there are small flocks across the country.

Description

Black face and legs, with short ears. There is a tuft of crew-cut wool on the forehead; the fleece is white and the legs are clean. The rams and ewes are without horns.

The Llanwenog (pronounced Thlan–wen-og) can trace its origins to the late 1800s in west Wales. The sheep will thrive on the harsh conditions 1,000 feet above sea level but can also take advantage of lowland pastures. The Llanwenog is easier to shepherd than most Welsh sheep having little desire to wander. They live longer than many other breeds and produce many lambs. The wool is very soft to the touch and is used for fine tweeds, hosiery, flannel and knitting-wool. The Llanwenog also produces top-quality meat.

Lleyn

Location

British Isles: originally Wales and the Welsh borders. Now found in all parts of Britain and Ireland.

Description

A medium-sized sheep with white face and head with black nostrils and often with black spots on the ears; no wool on face and legs. The rams and ewes are without horns.

The Lleyn (pronounced kleen) sheep is named after the Lleyn peninsula in North Wales. At the turn of the nineteenth century, milk from the Lleyn was used for cheese-making. In the 1970s a few breeders formed the Lleyn breed society to save the breed from extinction. The very fine wool is used for hand-knitting, hosiery and dress fabrics. The Lleyn also produces quality meat.

Lonk

Location

British Isles: originally east Lancashire, west Yorkshire and north Derbyshire, but now found country-wide.

Description

Pure black-and-white, wool-less face and legs. The rams and ewes have curled horns.

The Lonk has been bred on the Pennines from time immemorial; a flock in Lancashire can be traced back to 1740 when the monks of Sawley and Whalley farmed them on the Rossendale hills, near Blackburn. The Lonk is very hardy and lives throughout the year on the poor grazing provided by bleak moors between 1,000 and 2,000 feet above sea level. The finest grades of Lonk wool are used for hand-knitting wools and blankets. The coarse wool is used for carpets, rugs and tweeds. The Lonk also produces top-quality meat.

25.

Masham

Description

A distinctive black-and-white face and legs, and a top knot. The rams and ewes are without horns.

Masham (pronounced Mass-ham) sheep have been bred for over a hundred years in the northern counties of England, and are known by those who attempt to shear them as the 'mad' Masham. The Masham's hardiness and longevity are a result of its parentage: a cross of the Teeswater or Wensleydale ram with a Dalesbred or Swaledale ewe. The lustrous fleece is in great demand for speciality uses including the Italian fashion industry, and it is also used for quality carpets and upholstery. The Masham produces quality meat.

Location

British Isles: mainly the north of England.

North Country Cheviot

Location

British Isles: originally the Scottish Cheviots, then Caithness, Sutherland and Ross-shire, but now country-wide. It is also found in North America.

Description

A large, long, white sheep. The white face and legs are free of wool. Although the ewes are without horns, some rams may have them.

One of the largest hill sheep, the North Country Cheviot has been grazing the border Cheviot Hills from the mid-eighteenth century. In the late eighteenth century large numbers were driven to the north and west of Scotland where they thrived on vast tracts of land depopulated in the highland clearances. The hard-wearing wool is used for characteristic Scottish products such as tweeds, sports jackets, overcoats, travel rugs and carpets. The North Country Cheviot also produces top-quality meat.

Oxford Down

Location

British Isles: originally the Cotswolds, Midlands and south Wales as well as areas of Yorkshire and Aberdeenshire. Now found throughout the country and all over the world.

Description

A large sheep with a dark brown head with a wool top knot and wool on the cheeks. It has long ears and dark legs. The rams and ewes are without horns.

Originated in 1830 by crossing Cotswold rams with Hampshire Down and Southdown ewes, the Oxford Down is the largest of the British down breeds. The wool is used for hand-knitting yarns and hosiery. The Oxford Down also produces top-quality meat.

Hill Radnor

Description

A grey aquiline nose, and light brown face and legs which are free of wool. The rams have long, curved horns spiraling outwards. The ewes are without horns.

The Hill Radnor breed was developed in Radnorshire in a period beyond living memory. It is a hardy breed which can look after itself in both lowland and mountains in the worst weather conditions. The wool is used for high-quality fabrics such as fine tweeds and soft flannel and is in great demand by local hand-spinners and weavers. The Hill Radnor also produces top-quality meat.

Location

British Isles: mainly the hills bordering Radnor, Hereford, Brecon and Monmouthshire.

Romney

Location

British Isles: originally Kent and east Sussex, though there are now flocks country-wide. It is also found in North America, New Zealand and the Falkland Islands.

Description

A large longwool sheep with a broad, white face and a black nose. Some have a woolly top knot. The rams and ewes are without horns.

Seen on the Romney and north Kent marshes for eight hundred years. The Romney has always been prized for its wool, way back to the wool-smuggling days of the 1600s. The Romney is probably the most numerous breed in the world. Their feet are hard and strong to cope with the Kent marshes. The Romney has a habit of spreading out evenly over the grazing area to make the best of the pasture. The wool is used for high-quality rugs, carpets and hosiery as well as hand-knitting wools, worsted cloths, woollen cloth, baby clothes and blankets. The Romney also produces top-quality meat.

Rough Fell

Location

British Isles: although they are mostly on the Cumbrian and north Pennine fells there are small flocks across the country. It is also found in Europe.

Description

A black head with a large white patch on the nose. The legs have irregular black-and-white markings. The rams and ewes have strong, curled horns.

The Rough Fell is a descendant of a breed found in northern Britain in the middle ages. There is a reference to the breed at a show in the Yorkshire Dales in 1848. The Rough Fell is hardy and survives in the harsh conditions of the Cumbrian and Pennine fells. The toughness and springiness of the Rough Fell fleece makes it ideal for carpet manufacture, and it is also used in the Italian mattress market. The Rough Fell also produces quality meat.

Ryeland

Location

British Isles: across the country. It is also found in Australia and New Zealand.

Description

Dull, white face and legs well covered with wool. The rams and ewes are without horns.

This is probably the oldest recognised British sheep breed, originating over eight hundred years ago. Six hundred years ago the monks of Leominster were keeping the breed on the rye-growing district of Herefordshire. The Ryeland is among the minority breeds under the umbrella of the Rare Breeds Survival Trust. Today the wool is used for high-quality tweeds, hand-knitting wools and hosiery. The Ryeland also produces top-quality meat.

Shetland

Location

British Isles: originally the Shetland Isles but now country-wide. Also bred in North America.

Description

The Shetland has a small body, erect ears and bright eyes. Although the wool is mainly white there are small numbers of red, grey-brown and fawn sheep. The rams have beautifully rounded horns and the ewes are without horns.

The Shetland has developed in relative isolation since its possible introduction by the Vikings in the late eighth century. The islanders developed the breed to produce the finest wool of all indigenous British sheep, used for the famous Fair Isle sweaters, stockings and tweed. Although the white wool fetches the highest price, the coloured fleeces are very popular with hand-spinners. The red goes by the Shetland name of 'moorit'. The Shetland also produces high-quality meat.

Shropshire

Location

British Isles: originally from Shropshire and Staffordshire but now country-wide. The breed is also found in Europe and North America.

Description

A clean, soft, black face with a covering of wool on the top of the head. The legs are set well apart and are soft black in colour. The rams and ewes are without horns.

Developed in the nineteenth century by improving the local Shropshire and Staffordshire sheep. These sheep are bred to survive the demands of the Shropshire and Staffordshire hills. Shropshire sheep can adapt themselves to many types of pasture including the very sparse. In Scotland, Christmas tree growers use the Shropshire to control ground weeds. These sheep have a greater covering of wool than similar breeds, the fleece being dense, heavy and soft to handle. The wool is used for hosiery, hand-knitting wools and worsted suitings; large quantities are exported to Europe. The Shropshire is a source of top-quality meat.

Southdown

Location

British Isles: country-wide. Also in America, New Zealand and Australia.

Description

Light brown to mousey-coloured face and legs with the top of the head and ears covered with short wool, almost crew cut.

The Southdown has been on the Sussex Downs for many centuries. About two hundred years ago John Ellman started improving the breed by selection from within the flock and this was the foundation of the New Zealand 'Canterbury lamb'. The Southdown's excellent wool is used in the manufacture of high-quality fabrics, hosiery and hand-knitting wools. The Southdown also produces high-quality meat.

Suffolk

Description

Distinctive all-black head, ears and legs. The ears are carried at right angles to the head. The rams and ewes are without horns.

The Suffolk has been recognised as a pure breed since 1810; the first recorded mention of a Suffolk was in 1797. In medieval times the wool from shortwool sheep similar to the Suffolk was highly valued. The fleece is used for hosiery, hand-knitting yarns and tweeds. The Suffolk also produces good-quality meat.

Location

British Isles: across the country. Also found in the Netherlands, Europe, Australia, America and New Zealand

Swaledale

Location

British Isles: mainly on the Pennines in the northern counties of England.

Description

A dark head with a grey muzzle and short, strong hair on the face. Swaledale rams and ewes have horns which, set low, are round and rather wide. Swaledales are similar to Dalesbred sheep, but the Dalesbred have a speckled face.

Although the breed was first registered in 1919 and there is very little history recorded before this, farmers in North Yorkshire and Westmorland had specialised for generations in breeding similar sheep. The Swaledale is very hardy and able to survive being exposed on the high moorlands. The fine-quality wool is used for tweeds, rug wool and the thicker hand-knitting yarns. The coarser wool is used for carpet manufacture. The Swaledale also produces top-quality meat.

Teeswater

Description

The Teeswater has a pale, grey face with a top knot and ringlets with brown, mottled markings around the eyes and nose. The Teeswater has no wool on its face or legs below the knees. The rams and ewes are without horns.

The Teeswater is indigenous to Teesdale in County Durham and has been bred in the area for two hundred years, first being mentioned in 1798. The Teeswater has been cross-bred with the Swaledale, Dalesbred and Rough Fell, producing the Masham. There are many uses for the wool including tweeds, lining material and carpets. The Teeswater also produces top-quality meat.

Location

British Isles: originally the north of England but now country-wide. Also found in North America.

Welsh Mountain Badger Face Torddu

Location

British Isles: mostly Wales with small flocks throughout the country.

Description

White, grey or light brown with distinct badger stripes above the eyes. A black band runs from jaw to belly to the tip of the tail, and there are tan stripes on the black legs. The rams have dark, spiral horns while the ewes are without horns.

The Torddu, the Welsh name meaning 'black belly', is a very ancient breed of hardy mountain sheep mentioned in the Domesday Book. The Torddu became less popular in the middle ages when wool traders demanded white wool. Today the majority of wool is used for carpets. Torddu ewes are excellent mothers. The Torddu is primarily a meat breed producing delectable lamb on any grazing.

Welsh Mountain Badger Face Torwen

Location

British Isles: mostly Wales with small flocks throughout the country.

Description

Black, with a distinct white badger stripe above the eyes which is smaller than the eye stripe of the Torddu. There is white under the belly and the tail. The legs are tan with a black stripe. Rams have dark, spiral horns, while the ewes are without horns.

The Torwen, the Welsh name meaning 'white belly', is a very ancient breed of hardy mountain sheep, mentioned in the Domesday Book. They are less numerous than the Torddu sheep. The Torwen are excellent mothers. Primarily a meat breed, they produce delicious lamb from any grazing. The majority of the wool is used for carpets.

Description

Wensleydale

The largest British sheep, with a dark blue face and long, curly fleece and forelock.

The Wensleydale was developed from a ram called 'Bluecap', born in 1839 in a small hamlet near Bedale in North Yorkshire. Bluecap's parents were a Dishley ram and longwool ewe of a type that is now extinct. The soft Wensleydale wool – the finest lustre wool in the world – is used for high-quality knitting and weaving yarns. The Wensleydale is also a source of top-quality meat.

Location

British Isles: originally from north Yorkshire but now country-wide. Also found in Europe and North America.

Whiteface Dartmoor

Description

A white head and face, the face of the ewe being free of wool. The ears are short and thick with occasional black spots on them. The wool is white and has a fairly strong curl. The ram may have horns.

The Whiteface Dartmoor is descended from native sheep that have grazed on the moors since the seventeenth century. The Whiteface Dartmoor is very hardy and can survive on very poor pasture, grazing at altitudes from 500 to 2,000 feet above sea level. It is kept on the moor from May to December. The wool is mainly used for carpets. The Whiteface Dartmoor also produces top-quality meat.

Location

British Isles:
South-west England.

Sheep Talk

There are many words associated with sheep, both regional and national, which to those not involved in farming are a foreign language. Here are just a few of those terms:

Ewe	Female sheep one year old or older
Ram	Male sheep one year old or older
Lamb	A young sheep not yet weaned
Polled	A sheep that is naturally without horns
Tupping	Sheep sex
Couples	Ewes and lambs
Cade lambs	Hand-reared lambs
Lugmark	Ear-tag
Maid	Barren ewe
Clarts	Dung attached to fleece.

Lowland sheep are bred to live on lush, enclosed pastures so they tend to be heavier than their hill-sheep cousins. They give birth to one or two lambs from Christmas onwards.

Hill and mountain sheep are bred to survive in the harsh upland terrain where they graze the hills and fells with little or no restriction. Because of the poor and less nutritious grazing the sheep are lighter. They give birth to just one lamb in April or May.